Bluetooth Low Energy: A Technical Primer

1st Edition

Tony Gaitatzis

BackupBrain Publishing, 2017

ISBN: 978-1-7751280-8-3

backupbrain.co

Bluetooth Low Energy: A Technical Primer

by Tony Gaitatzis

backupbrain@gmail.com

(This page intentionally left blank)

Dedication

To Shade, for that chicken soup

and Andrew, for being a sounding board

(This page intentionally left blank)

Preface

Thank you for buying this book. I'm excited to have written it and more excited that you are reading it.

I started with Bluetooth Low Energy in 2011 while making portable brain imaging technology. Later, while working on a friend's wearable electronics startup, I ended up working behind teh scenes on the TV show America's Greatest Makers in the Spring of 2016.

Coming from a web programming background, I found the mechanics and nomenclature of BLE confusing and cryptic. A er immersing myself in it for a period of time I acclimated to the di erences and began to appreciate the power behind this low-power technology.

Unlike other wireless technologies, BLE can be powered from a coin cell battery for months at a time - perfect for a wearable or Internet of Things (IoT) project! Because of its low power and short data transmissions, it is great for transmitting bite size information, but not great for streaming data such as sound or video.

Good luck and enjoy!

Introduction

In this book, you will learn the mechanics behind the Bluetooth Low Energy protocol, a wireless technology that lets computers, smartphones, and smart devices communicate within a space about the size of a room.

Through the course of the book you will learn important concepts that relate to:

- How Bluetooth Low Energy works,
- How data is sent and received
- Common paradigms for handling data

Additionally, you will learn how three common product paradigms work:

- An iBeacon
- An Echo Server
- A Remote Controlled Device

This book is an excellent read for anyone who wants to know the terminology, technology, and concepts behind Bluetooth Low Energy devices but doesn't need to program them.

Overview

Bluetooth Low Energy (BLE) is a digital radio protocol. Very simply, it works by transmitting radio signals from one computer to another.

Bluetooth supports a hub-and-spoke model of connectivity. One device acts as a hub, or “Central” in Bluetooth terminology. Other devices act as “Peripherals.”

A Central may hold several simultaneous connections with a number of peripherals, but a peripheral may only hold one connection at a time (Figure 1-1). Hence the names Central and Peripheral.

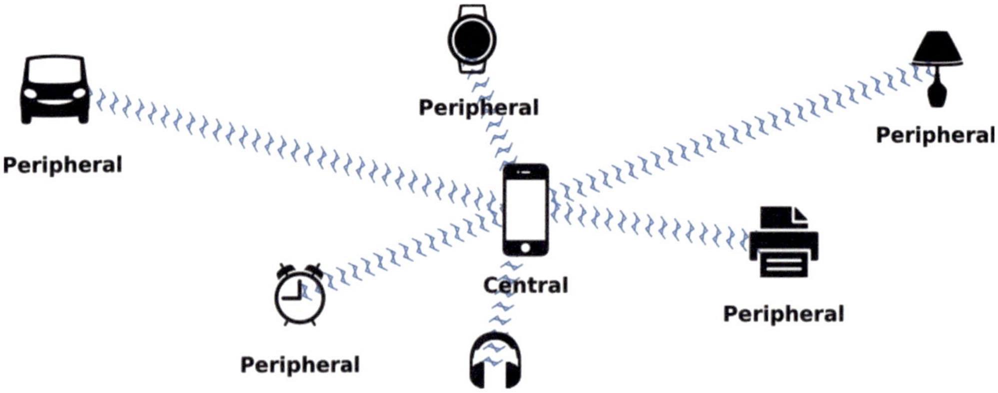

Figure 1-1. Bluetooth network topology

For example, your smartphone acts as a Central. It may connect to a Bluetooth speaker, lamp, smartwatch, and fitness tracker. Your fitness tracker and speaker, both Peripherals, can only be connected to one smartphone at a time.

The Central has two modes: scanning and connected. The Peripheral has two modes: advertising and connected. The Peripheral must be advertising for the Central to see it.

Advertising

A Peripheral advertises by advertising its device name and other information on one radio frequency, then on another in a process known as frequency hopping. In doing so, it reduces radio interference created from reflected signals or other devices.

Scanning

Similarly, the Central listens for a server's advertisement first on one radio frequency, then on another until it discovers an advertisement from a Peripheral. The process is not unlike that of trying to find a good show to watch on TV.

The time between radio frequency hops of the scanning Central happens at a different speed than the frequency hops of the advertising Peripheral. That way the scan and advertisement will eventually overlap so that the two can connect.

Each device has a unique media access control address (MAC address) that identifies it on the network. Peripherals advertise this MAC address along with other information about the Peripheral's settings.

Connecting

A Central may connect to a Peripheral after the Central has seen the Peripheral's advertisement. The connection involves some kind of handshaking which is handled by the devices at the hardware or firmware level.

While connected, the Peripheral may not connect to any other device.

Disconnecting

A Central may disconnect from a Peripheral at any time. The Peripheral is aware of the disconnection.

Communication

A Central may send and request data to a Peripheral through something called a "Characteristic." Characteristics are provided by the Peripheral for the Central to access. A Characteristic may have one or more properties, for example READ or WRITE. Each Characteristic belongs to a Service, which is like a container for Characteristics. This paradigm is called the Bluetooth Generic Attribute Profile (GATT).

The GATT paradigm is laid out as follows (Figure 1-2).

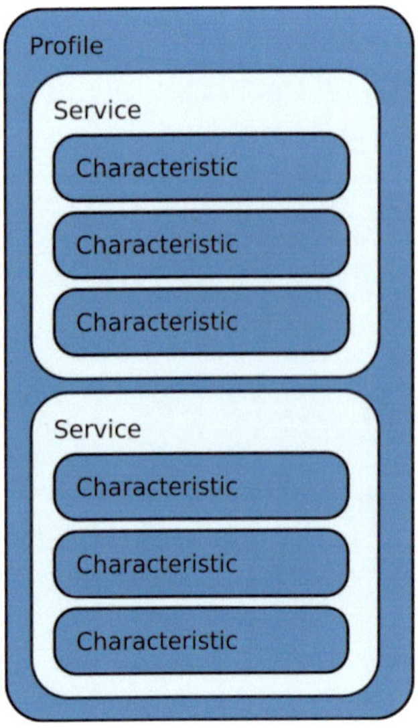

Figure 1-2. Example GATT Structure

To transmit or request data from a Characteristic, a Central must first connect to the Characteristic's Service.

For example, a heart rate monitor might have the following GATT profile, allowing a Central to read the beats per minute, name, and battery life of the server (Figure 1-3).

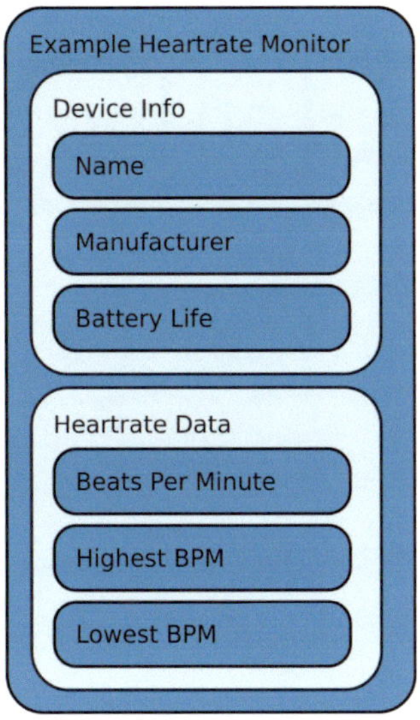

Figure 1-3. Example GATT structure for a heart monitor

In order to retrieve the battery life of the Characteristic, the Central must be connected also to the Peripheral's "Device Info" Service.

Because a Characteristic is provided by a Peripheral, the terminology refers to what can be done to the Characteristic. A "write" occurs when data is sent to the Characteristic and a "read" occurs when data is downloaded from the Characteristic.

To reiterate, a Characteristic is a field that can be written to or read from. A Service is a container that may hold one or more Characteristics. GATT is the layout of these Services and Characteristics. Characteristic can be written to or read from.

Byte Order

Bluetooth orders data in both Big-Endian and Little-Endian depending on the context.

During advertisement, data is transmitted in Big Endian, with the most significant bytes of a number at the end (Figure 1-4).

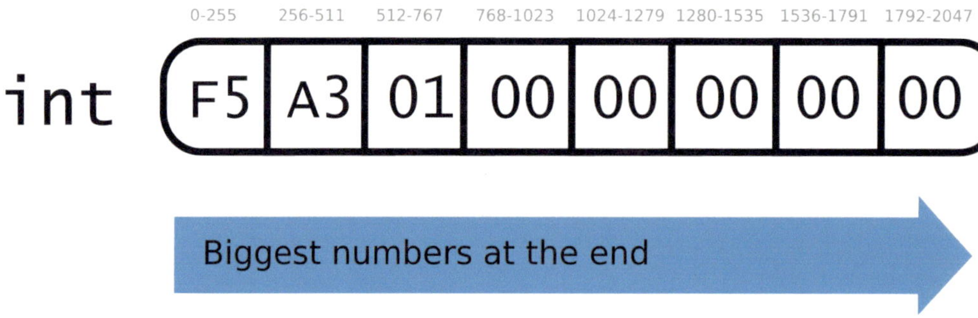

Figure 1-4. Big Endian byte order

Data transfers inside the GATT however are transmitted in Little Endian, with the least significant byte at the end (Figure 1-5).

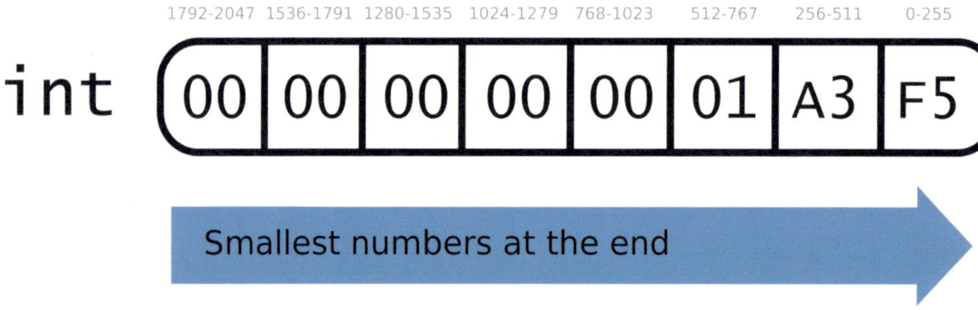

Figure 1-5. Little Endian byte order

Permissions

A Characteristic grants certain Permissions of the Central. These permissions include the ability to read and write data on the Characteristic, and to subscribe to Notifications.

Descriptors

Descriptors describe the configuration of a Characteristic. The only one that has been specified so far is the "Notification" flag, which lets a Central subscribe to Notifications.

UUIDs

A UUID, or Universally Unique IDentifier is a very long identifier that is likely to be unique, no matter when the UUID was created or who created it.

BLE uses UUIDs to label Services and Characteristics so that Services and Characteristics can be identified accurately even when switching devices or when several Characteristics share the same name.

For example, if a Peripheral has two "Temperature" Characteristics - one for Celsius and the other in Fahrenheit, UUIDs allow for the right data to be communicated.

UUIDs are usually 128-bit strings and look like this:

```
ca06ea56-9f42-4fc3-8b75-e31212c97123
```

But since BLE has very limited data transmission, 16-bit UUIDs are also supported and can look like this:

```
0x1815
```

Each Characteristic and each Service is identified by its own UUID. Certain UUIDs are reserved for specific purposes.

For example, UUID 0x180F is reserved for Services that contain battery reporting Characteristics.

Similarly, Characteristics have reserved UUIDs in the Bluetooth Specification.

For example, UUID 0x2A19 is reserved for Characteristics that report battery levels.

A list of UUIDs reserved for specific Services can be found in ***Appendix IV: Reserved GATT Services.***

A list of UUIDs reserved for specific Characteristics can be in ***Appendix V: Reserved GATT Characteristics.***

If you are unsure what UUIDs to use for a project, you are safe to choose an unassigned service (e.g. 0x180C) for a Service and generic Characteristic (0x2A56).

Although the possibility of two generated UUIDs being the same are extremely low, programmers are free to arbitrarily define UUIDs which may already exist. So long as the UUIDs defining the Services and Characteristics do not overlap in the a single GATT Profile, there is no issue in using UUIDs that exist in other contexts.

Bluetooth Hardware

All Bluetooth devices feature at least a processor and an antenna (Figure 1-6).

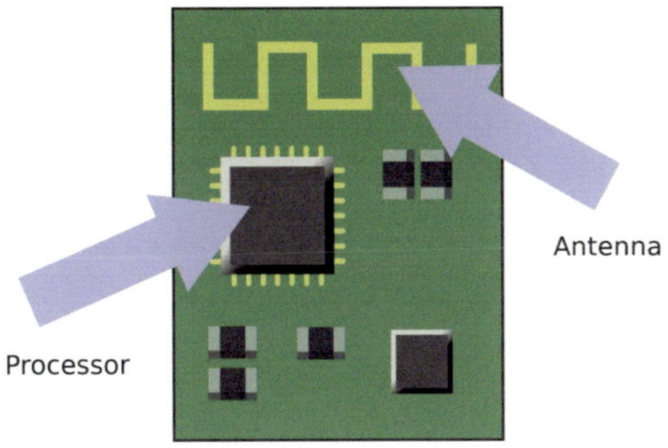

Figure 1-6. Parts of a Bluetooth device

The antenna transmits and receives radio signals. The processor responds to changes from the antenna and controls the antenna's tuning, the advertisement message, scanning, and data transmission of the BLE device.

Power and Range

BLE has 20x2 Mhz channels, with a maximum 10 mW transmission power, 20 byte packet size, and 1 Mbit/s speed. This means it's great for short-range or ad-hoc networks such as those inside a house, around a person's body, or as people or devices coming in and out of range.

As with any radio signal, the quality of the signal drops dramatically with distance, as shown below (Figure 1-7).

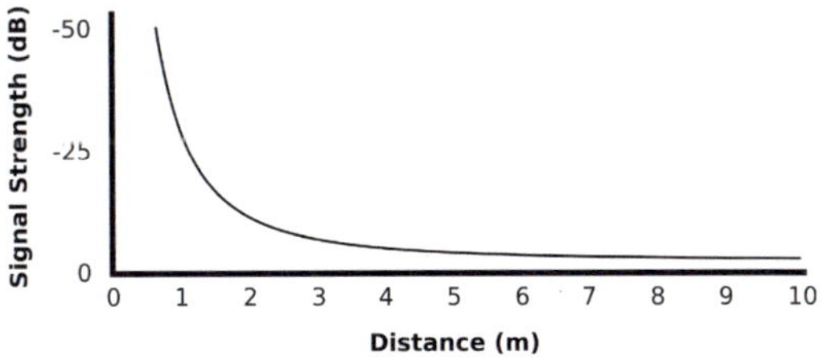

Figure 1-7. Distance versus Bluetooth Signal Strength

This signal quality is correlated the Received Signal Strength Indicator (RSSI).

If the RSSI is known when the Peripheral and Central are 1 meter apart (A), as well as the RSSI at the current distance (R) and the radio propagation constant (n). The distance betweeen the Central and the Peripheral in meters (d) can be approximated with this equation:

$$d \approx 10^{\frac{A-R}{10n}}$$

The radio propagation constant depends on the environment, but it is typically somewhere between 2.7 in a poor environment and 4.3 in an ideal environment.

Take for example a device with an RSSI of 75 at one meter, a current RSSI reading 35, with a propagation constant of 3.5:

$$d \approx 10^{\frac{75-35}{10 \times 3.5}}$$

$$d \approx 10^{\frac{40}{35}}$$

$$d \approx 14$$

Therefore the distance between the Peripheral and Central is approximately 14 meters.

Data Length and Speed

It is worth noting that Bluetooth Low Energy has a maximum data packet size of 20 bytes, with a 1 Mbit/s speed. This is slow when compared to protocols such as WiFi or 4G, with speeds of 300 Mbit/s and 10 Mbit/s respectively.

Power Consumption

Bluetooth Low Energy also consumes dramatically less power - 2 milliwatts compared to WiFi's 80 milliwatts or 4G/LTE's 1000 milliwats. The result is that a Bluetooth Low Energy device can run on a coin cell battery for months or even years, but a WiFi or 4G/LTE device will die after a few minutes on the same battery.

Scanning and Advertising

The first step to any Bluetooth Low Energy interaction is for the Peripheral to make the Central aware of its existence, through a process called Advertising.

Advertising reports the server name and other information one channel at a time until there are no more channels and the server repeats the process again at the first channel.

During the Advertising process, a Peripheral Advertises while a Central scans. A Peripheral may start or stop advertising at any time.

Bluetooth devices discover each other when they are tuned to the same radio frequency, also known as a Channel. There are three channels dedicated to device discovery in Bluetooth Low Energy: (Table 2-1):

Table 2-1. Bluetooth Low Energy Discovery Radio Channels

Channel	Radio Frequency
37	2402 Mhz
39	2426 Mhz
39	2480 Mhz

The peripheral will advertise its name and other data over one channel and then another. This is called Frequency Hopping (Figure 2-1).

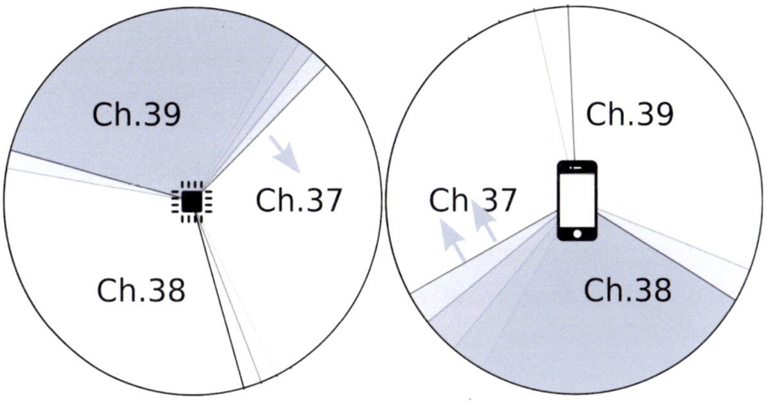

Figure 2-1. Advertise and scan processes

Similarly, the Central listens for advertisements first on one channel and then another. The Central hops frequencies faster than the Peripheral, so that the two are guaranteed to be on the same channel eventually.

A Peripheral may advertise from 100ms to 100 seconds depending on its configuration, changing channels every 0.625ms (Figure 2-2).

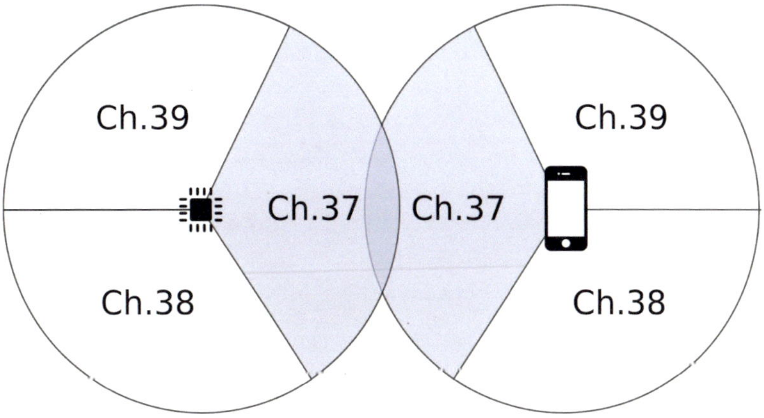

Figure 2-2. Scan finds Advertiser

A Central's Scanning settings vary wildly, for example scanning every 10ms for 100ms, or scanning for 1 second for 10 seconds. A shorter scan period with a greater consumes less energy, but is slower to find nearby Peripherals.

Typically when a Central discovers advertising Peripheral, the Central requests a Scan Response from the Peripheral. In some cases, the Scan Response contains useful data. For example, iBeacons use Scan Response data to inform Centrals of each iBeacon's location without the Central needing to connect and download more data.

Example

An example of this is an app on iPhone that scans for nearby Bluetooth Low Energy Peripherals (Figure 2-3).

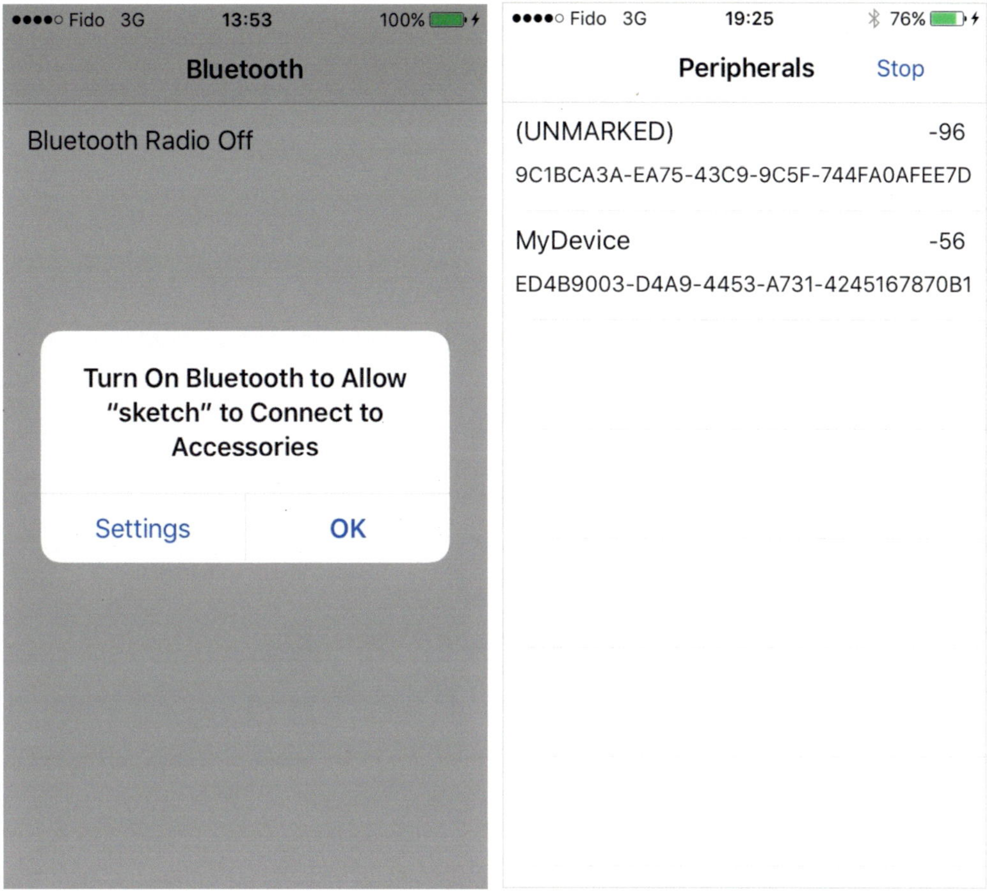

Figure 2-3. iPhone asks for permission to turn on Bluetooth radio and scans for nearby Bluetooth Low Energy Peripherals

Connecting

Once a Central has discovered a Peripheral, the central can attempt to connect. This must be done before data can be passed between the Central and Peripheral. A Central may hold several simultaneous connections with a number of peripherals, but a Peripheral may only hold one connection at a time. Hence the names Central and Peripheral (Figure 3-1).

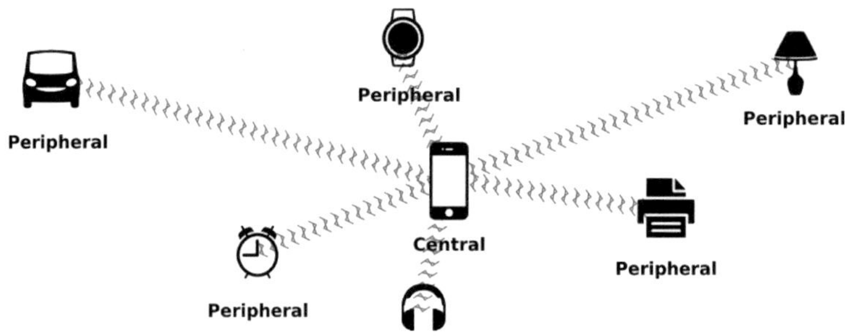

Figure 3-1. Bluetooth network topology

Bluetooth supports data 37 data channels ranging from 2404 MHz to 2478 MHz.

Once the connection is established, the Central and Peripheral negotiate which of these channels to begin communicating over. As part of this, a unique Media Access Control (MAC) address of the Central is sent to the Peripheral.

A MAC address is a 48-bit address given to every network device. It is typically represented in a hexadecimal format, similar to this:

```
08:00:27:0E:25:B8
```

Because the Peripheral can only hold one connection at a time, it must disconnect from the Central before a new connection can be made.

The connection and disconnection process works like this (Figure 3-2).

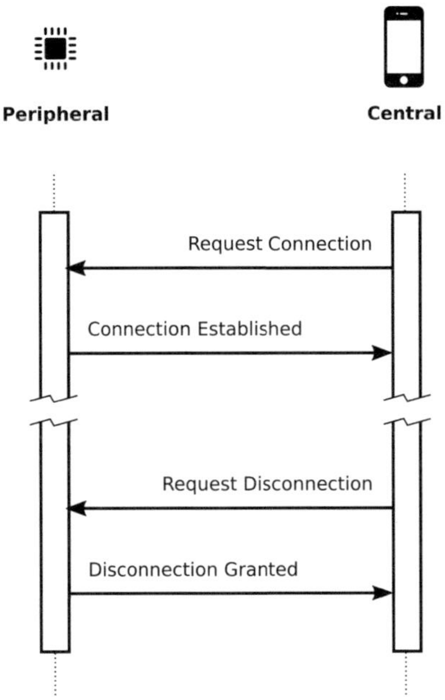

Figure 3-2. Connection and disconnection process

Examples

An iPhone can connect to a nearby Advertising Peripheral and display information about it (Figure 3-3).

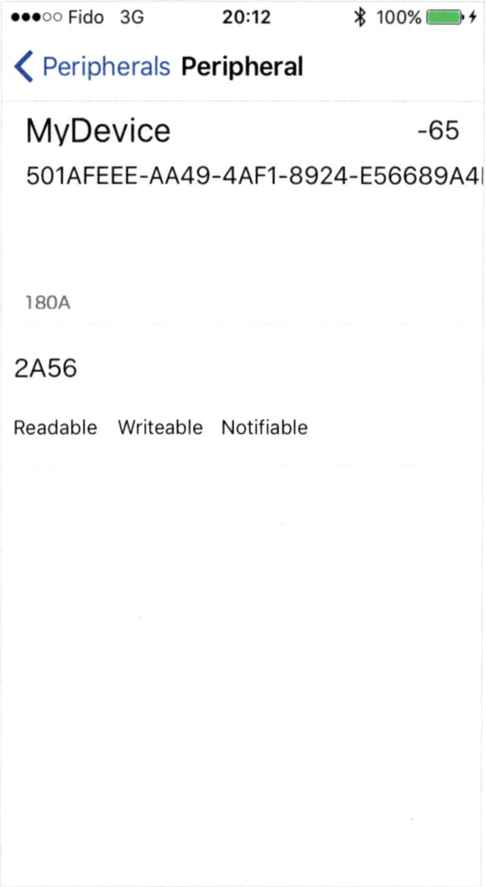

Figure 3-3. iPhone connects to a nearby Bluetooth Peripheral

The Peripheral in this example can can respond to a connection or disconnection by a Central. If the Peripheral were connected to some kind of debugging console, it may report each time a Central has connected or disconnected (Figure 3-4).

```
Central connected: 45:b5:7d:96:01:2f
Central disconnected
```

Figure35-4. Peripheral reporting via debugging console that a Central has connected, then disconnected.

Services and Characteristics

Before data can be transmitted back and forth between a Central and Peripheral, the Peripheral must host a GATT Profile. That is, the Peripheral must have Services and Characteristics.

Identifying Services and Characteristics

Each Service and Characteristic is identified by a Universally Unique Identifier (UUID). The UUID follows the pattern 0000XXXX-0000-1000-8000-00805f9b34fb, so that a 32-bit UUID 00002a56-0000-1000-8000-00805f9b34fb can be represented as 0x2a56.

Some UUIDs are reserved for specific use. For instance any Characteristic with the 16-bit UUID 0x2a35 (or the 32-bit UUID 00002a35-0000-1000-8000-00805f9b34fb) is implied to be a blood pressure reading.

For a list of reserved Service UUIDs, see ***Appendix IV: Reserved GATT Services***.

For a list of reserved Characteristic UUIDs, see ***Appendix V: Reserved GATT Characteristics***.

Generic Attribute Profile

Services and Characteristics describe a tree of data access points on the peripheral. The tree of Services and Characteristics is known as the Generic Attribute (GATT) Profile. It may be useful to think of the GATT as being similar to a folder and file tree (Figure 4-1).

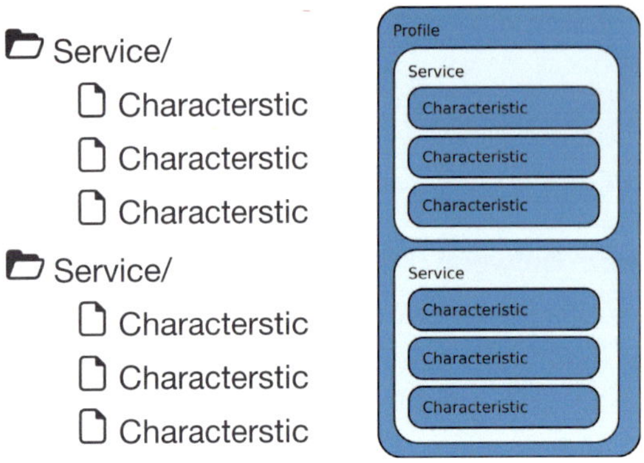

Figure 4-1. GATT Profile filesystem metaphor

Characteristics act as channels that can be communicated on, and Services act as containers for Characteristics. A top level Service is called a Primary service, and a Service that is within another Service is called a Secondary Service.

Permissions

Characteristics can be configured with the following attributes, which define what the Characteristic is capable of doing (Table 4-1):

Table 4-1. Characteristic Permissions

Descriptor	Description
Read	Central can read this Characteristic, Peripheral can set the value.
Write	Central can write to this Characteristic, Peripheral will be notified when the Characteristic value changes and Central will be notified when the write operation has occurred.
Notify	Central will be notified when Peripheral changes the value.

Because the GATT Profile is hosted on the Peripheral, the terms used to describe a Characteristic's permissions are relative to how the Peripheral accesses that Characteristic. Therefore, when a Central uploads data to the Peripheral, the Peripheral can "read" from the Characteristic. The Peripheral "writes" new data to the Characteristic, and can "notify" the Central that the data is altered.

Example

An Android app acting as a Central shows the GATT Profile of a connected Peripheral (Figure 4-2)

Figure 4-2. Android app shows GATT Profile of a connected Peripheral

Reading Data from a Peripheral

The real value of Bluetooth Low Energy is the ability to transmit data wirelessly.

Bluetooth Peripherals are passive, so they don't push data to a connected Central. Instead, Centrals make a request to read data from a Characteristic. This can only happen if the Characteristic enables the Read Attribute.

This is called "reading a value from a Characteristic."

Therefore, if a Peripheral changes the value of a Characteristic, then later a Central downloads data from the Peripheral, the process looks like this (Figure 5-1):

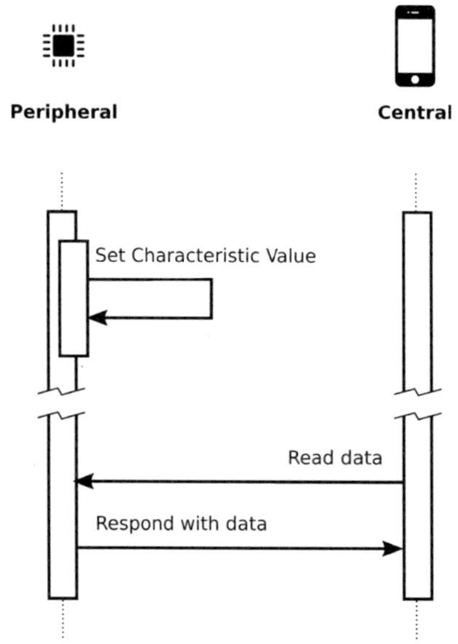

Figure 5-1. The process of a Central reading data from a Peripheral

A Central can read a Characteristic repeatedly, regardless if Characteristic's value has changed.

Examples

An iPhone app shows the Readable Permission of a Characteristic (Figure 5-2) and the data read from that Characteristic (Figure 5-3).

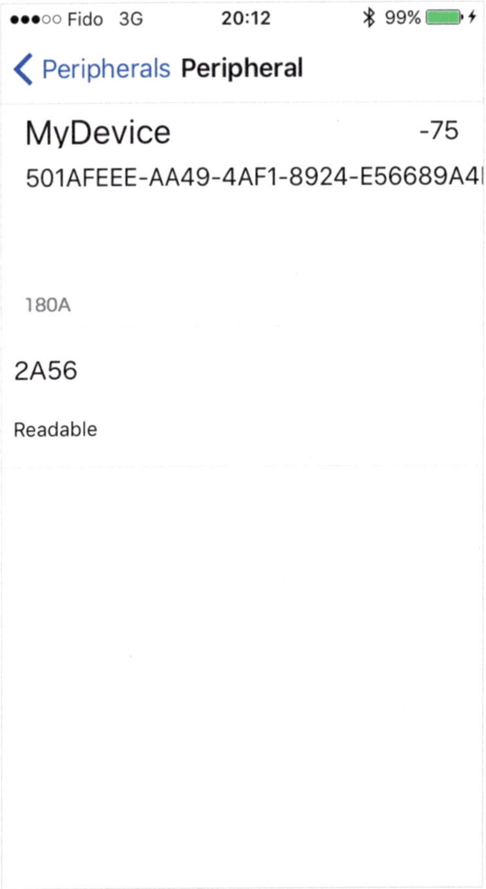

Figure 5-2. Readable Characteristic available on connected Peripheral

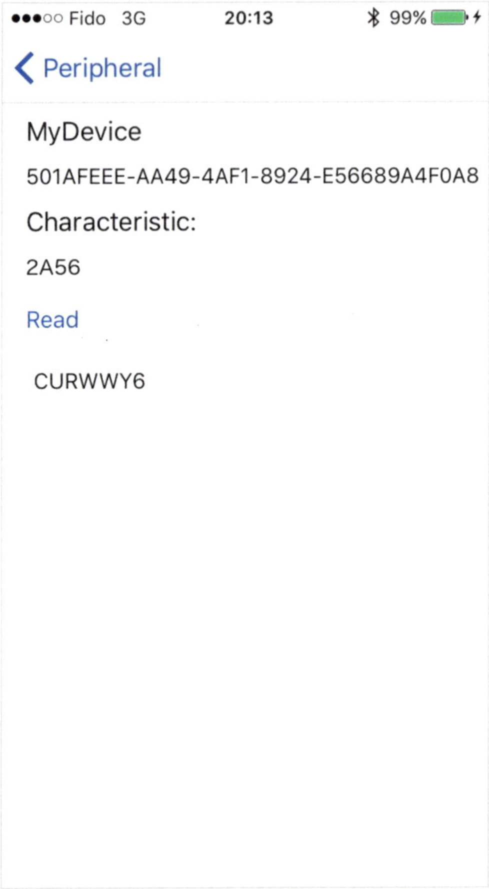

Figure 5-3. Data read from the Charactersitic of a connected Peripheral

The Peripheral in this example can change the value of one of its hosted Characteristics periodically. If the Peripheral were connected to some kind of debugging console, it may report each time the Central has written data, similar to this (Figure 5-4).

```
Setting characteristic to: HFVTh
Setting characteristic to: CURWWY6
Setting characteristic to: A6D7SZZFYQ8Sh
```

Figure 5-4. Peripheral reporting via debugging console that it has changed the data in one of its Characteristics.

Writing Data to a Peripheral

Data is sent from the Central to a Peripheral when the Central writes a value in a Characteristic hosted on the Peripheral, presuming that Characteristic has write permissions.

The process looks like this (Figure 6-1):

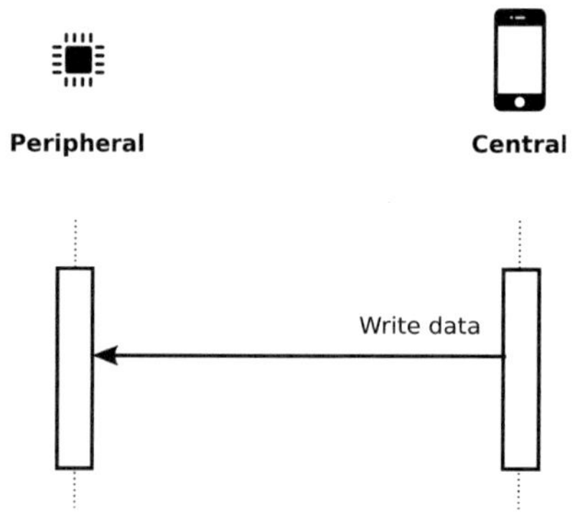

Figure 6-1. The process of a Central writing data to a Peripheral

Examples

An iPhone app acting as a Central shows the Readable Permission of a Characteristic on a connected Peripheral (Figure 6-2) and a text string queued to be sent to that Characteristic (Figure 6-3).

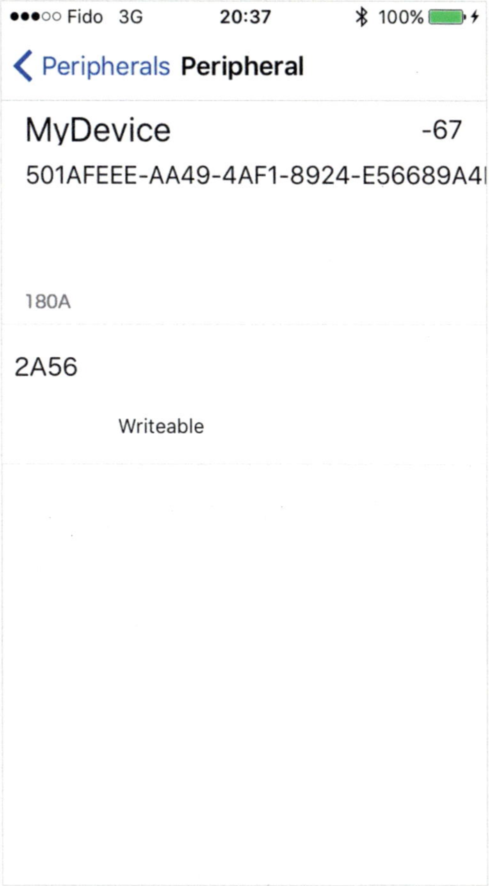

Figure 6-2. Writable Characteristic available on connected Peripheral

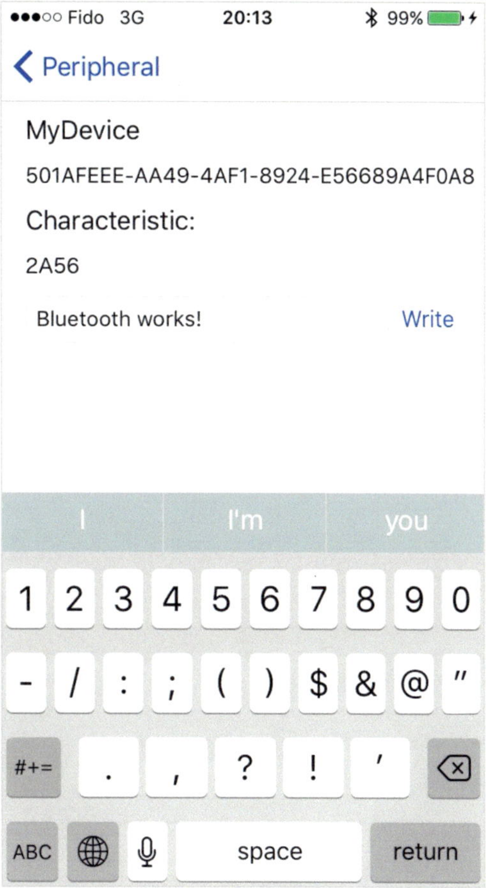

Figure 6-3. Data queued to be written to the Charactersitic of a connected Peripheral

The Peripheral in this example can read the newly written data on its Characteristic. If the Peripheral were connected to some kind of debugging console, it may report each time the Central has written data, similar to this (Figure 6-4).

```
5 bytes sent to characteristic 2A56: hello
5 bytes sent to characteristic 2A56: world
16 bytes sent to characteristic 2A56: Bluetooth works!
```

Figure 6-4. Peripheral reporting Characteristic Write events to a debugging console.

Using Notifications

Being able to read from the Central has limited value if the Central does not know when new data is available.

Notifications solve this problem. A Characteristic can issue a notification when it's value has changed. A Central that subscribes to these notifications will know when the Characteristic's value has changed, but not what that new value is. The Central can then read the latest data from the Characteristic.

The whole process looks something like this (Figure 7-1).

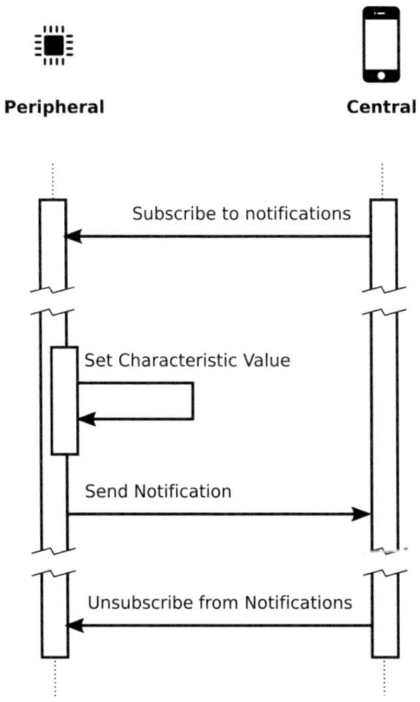

Figure 7-1. The process of a Peripheral notifying a connected Central of changes to a Characteristic

In order to support notifications, a Characteristics must have the Client Characteristic Configuration (0x2902) Descriptor, which must be writeable. Centrals can subscribe to notifications by setting the Descriptor value:

Table 7-1. Client Characteristic Configuration Descriptor values

Value	Description
0x0100	Enable notifications
0x0000	Disable notifications

Examples

An iPhone app acting as a Central shows the Notify Permission of a Characteristic on a connected Peripheral (Figure 7-2) and the data read from that Characteristic after Subscribing (Figure 7-3).

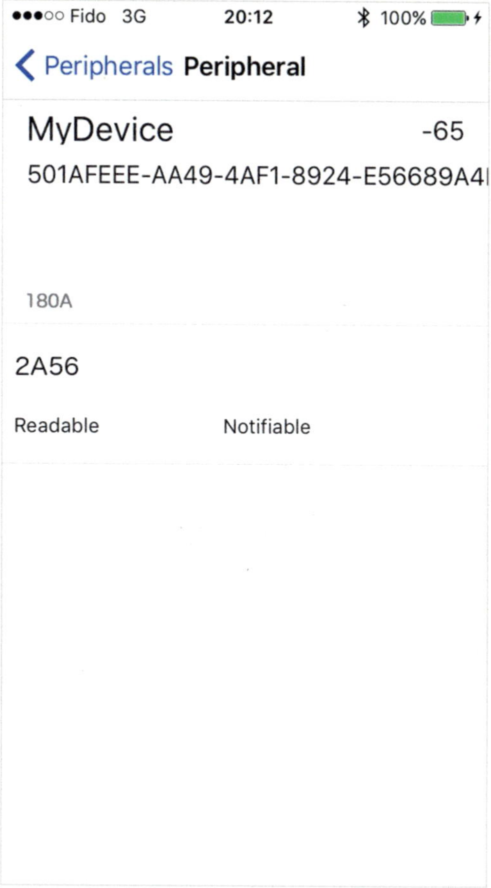

Figure 7-2. Writable Characteristic available on connected Peripheral

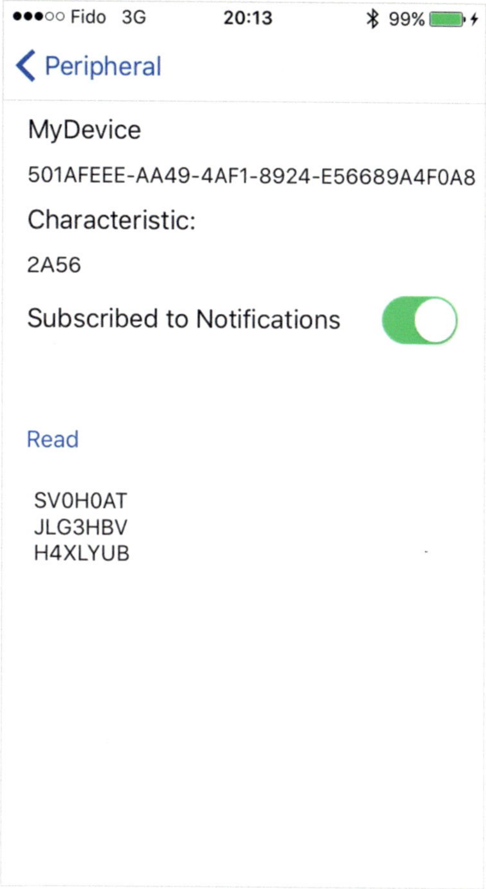

Figure 7-3. Data queued to be written to the Charactersitic of a connected Peripheral

The Peripheral in this example may set the Notifiable, Readable Characteristic to some random text string every few seconds. If the Peripheral were connected to some kind of debugging console, it may report something like this (Figure 7-4).

```
Setting characteristic to: SV0H0AT
Setting characteristic to: JLG3HBV
Setting characteristic to: H4XLYUB
```

Figure 7-4. Peripheral reporting that it has altered the content of one of its Characteristics.

Streaming Data

The maximum packet size you can send over Bluetooth Low Energy is 20 bytes. More data can be sent by dividing a message into packets of 20 bytes or smaller, and sending them one at a time

These packets can be sent at a certain speed.

Bluetooth Low Energy transmits at 1 Mb/s. Between the data transmission time and the time it may take for a Peripheral to process incoming data, there is a time delay between when one packet is sent and when the next one is ready to be sent.

To send several packets of data, a queue/notification system must be employed, which alerts the Central when the Peripheral is ready to receive the next packet.

There are many ways to do this. One way is to set up a Characteristic with read, write, and notify permissions, and to flag the Characteristic as "ready" after a write has been processed by the Peripheral. This sends a notification to the Central, which sends the next packet. That way, only one Characteristic is required for a single data transmission.

This process can be visualized like this (Figure 8-1).

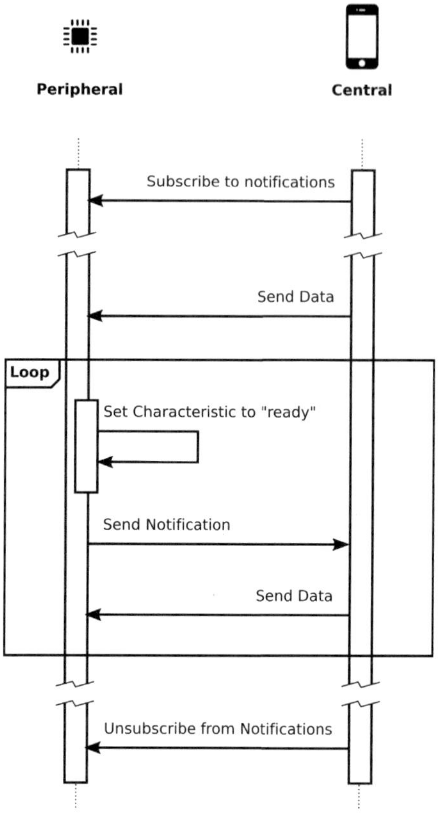

Figure 8-1. The process of using notifications to handle flow control on a multi-packed data transfer

The maximum packet size you can send over Bluetooth Low Energy is 20 bytes. More data can be sent by dividing a message into packets of 20 bytes or smaller, and sending them one at a time

These packets can be sent at a certain speed.

Bluetooth Low Energy transmits at 1 Mb/s. Between the data transmission time and the time it may take for a Peripheral to process incoming data, there is a time delay between when one packet is sent and when the next one is ready to be sent.

To send several packets of data, a queue/notification system must be employed, which alerts the Central when the Peripheral is ready to receive the next packet.

There are many ways to do this. One way is to set up a Characteristic with read, write, and notify permissions, and to flag the Characteristic as "ready" after a write has been processed by the Peripheral. This sends a notification to the Central, which sends the next packet. That way, only one Characteristic is required for a single data transmission.

Examples

An Android app acting as a Central shows the a Characteristic on a connected Peripheral with Read, Write, and Notify Permission (Figure 8-2) and the data read from that Characteristic after Subscribing (Figure 8-3).

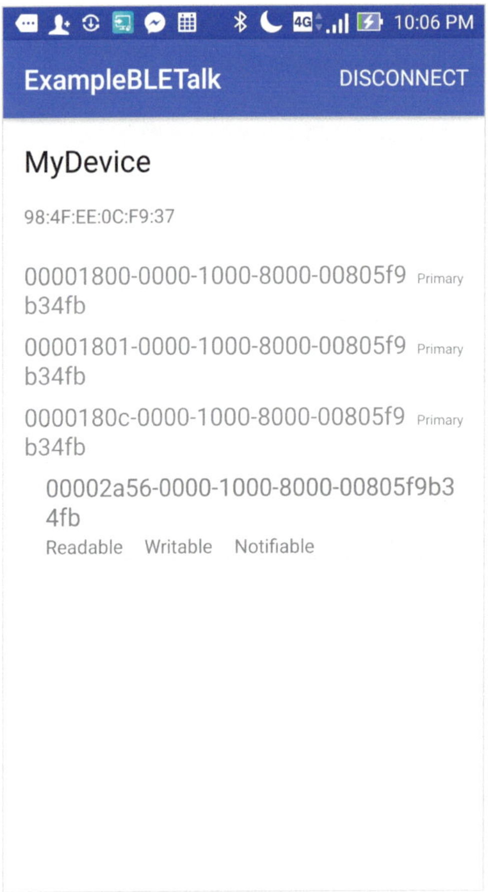

Figure 8-2. Read, Write, Notify Characteristic available on connected Peripheral

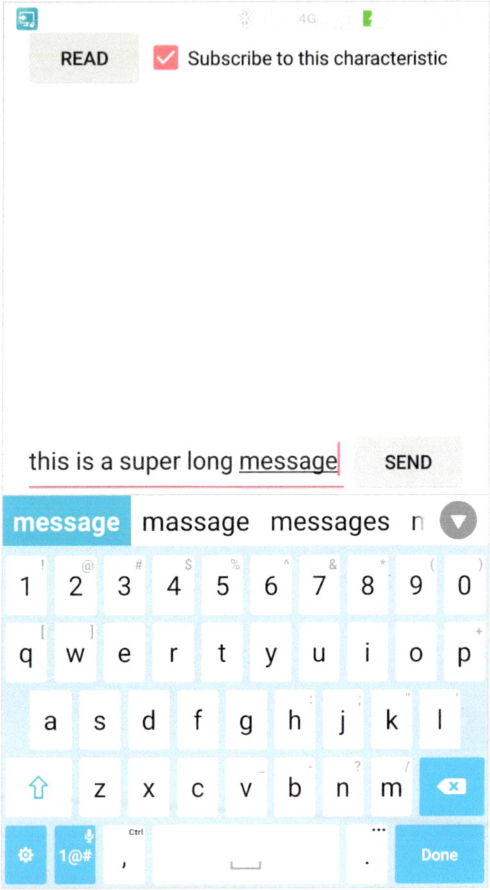

Figure 8-3. Data queued to be written to the Charactersitic of a connected Peripheral

The Peripheral in this example may process the inbound text behind the scenes and how that text is written to a Characteristic in chunks. If the Peripheral were connected to some kind of debugging console, it may report something like this (Figure 8-4).

```
16 bytes sent to characteristic 2A56: this is a super
Ready for more data
16 bytes sent to characteristic 2A56: long message
Ready for more data
```

Figure 8-4. Peripheral reporting what data a Characteristic has written to one of its Characteristic.

Example: iBeacons

Beacons can be used for range finding or spacial awareness. iBeacons are a special type of Beacon that is widely supported by the industry. It supports certain data that identifies the iBeacons to makes range finding and spacial awareness easier across platforms.

Due to the nature of how radio signals diminish in intensity with distance, Bluetooth Peripherals can be used both for range finding and spacial awareness.

Range Finding

Bluetooth signals can be used to approximate the distance between a Peripheral and a Central because the radio signal quality drops off in a predictable way with distance. The diagram below shows how the signal might drop as distance increases (Figure 9-1).

Figure 9-1. Distance versus Bluetooth Signal Strength

This drop-off rate, known as the Inverse-Square Law, is universal with electromagnetic radiation.

Due to radio interference and absorption from surrounding items, the radio signal propagation varies a lot from environment to environment, and even step to step. This makes it very difficult to know the precise distance between a Central and an iBeacon.

One or more Centrals can approximate their distance from a single iBeacon without connecting.

Spacial Awareness

A Central can approximate its position in space using a process called trilateration. Trilateration works by a computing series of equations when both the distance from and location of nearby iBeacons are known (Figure 9-2).

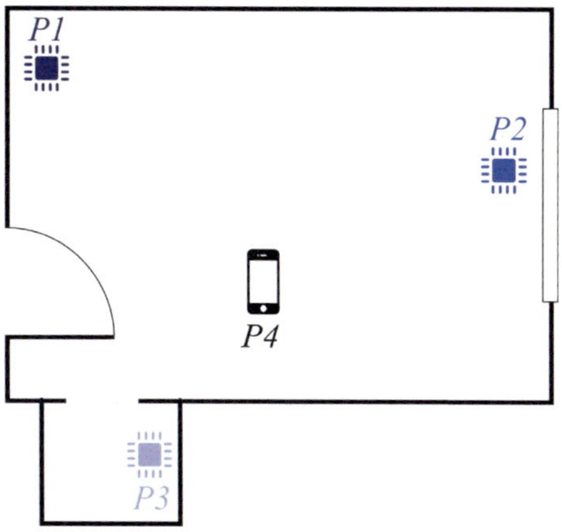

Figure 9-2. Example Central and iBeacon positions in a room

This is a pretty math-intensive process, but it's all based on the Pythagorus Theorem. By calculating the shape of the triangles made from the relative positions of all the iBeacons and the Central, one can determine the location of the Central (Figure 9-3).

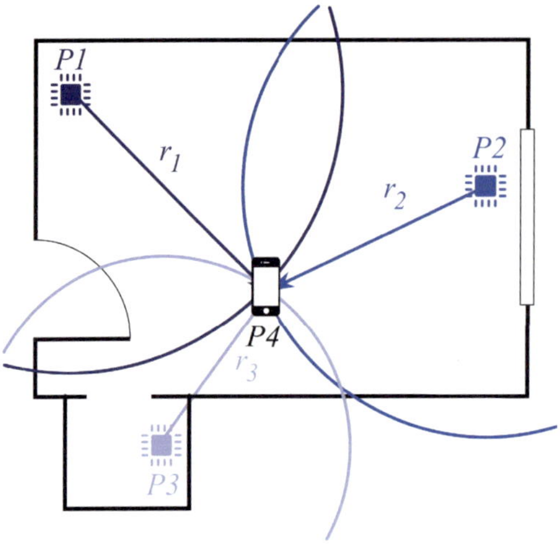

Figure 9-3. Distances from iBeacons to Central

iBeacons

The Scan Result allows a Central to read information from a Peripheral without connecting to it, in much the same way that the advertising name is read.

Although Android supports iBeacon scanning, it does not support iBeacon advertising. Therefore it is possible to build an app that discovers iBeacons but not possible to create an iBeacon in Android

iBeacons are beacons that advertise information about their location and advertise intensity using the Scan Result feature of Bluetooth Low Energy.

The Scan Result allows a Central to read information from a Peripheral without connecting to it, in much the same way that the Device name is advertised.

There are only two tricks to creating an iBeacon client in Android:

1. iBeacons are identified by their network (MAC) address instead of by name, because all iBeacons for the same location service have the same name.

2. The distance to the iBeacon is calculated based on the RSSI.

iBeacons do this by advertising certain data that can referenced when looking up where the iBeacons are located.

Table 9-1. iBeacon Advertised data

Data	Position	Length	Description
iBeacon Header	0	2	Identifies the Peripheral as an iBeacon
Manufacturer ID	6	2	A numeric value that represents the manufacturer
UUID	9	16	All iBeacons in a network have the same UUID
Major	25	2	A top level numeric identifier
Minor	27	2	A numeric identifier under the Major number
Transmission Power	29	1	Tells what the transmission power is in mDb

This data stored as binary in the Peripheral's advertising data. Unlike other data in Bluetooth Low Energy, numeric values in this packet are stored in big endian format (Figure 9-4).

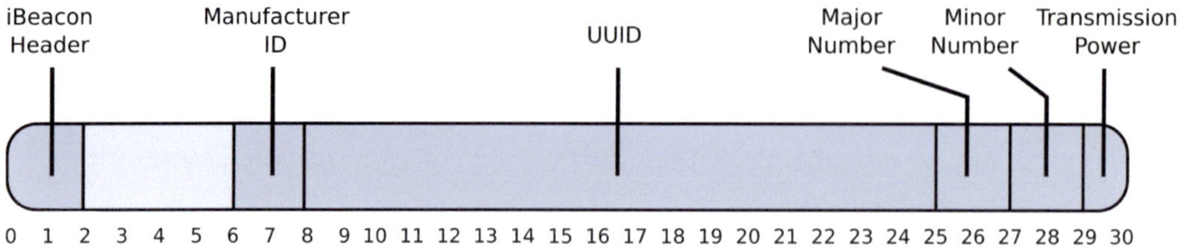

Figure 9-4. iBeacon Advertising Packet

An example implementation is a museum that has iBeacons at each exhibit in the museum. All iBeacons in the exhibit share the same UUID. The museum uses Major numbers to identify floors and Minor numbers to identify rooms of the exhibit.

The museum's smartphone app has an internal data set relating Major and Minor values to the floors and rooms of the exhibit. It scans for all iBeacons with a specific UUID. The nearby iBeacons are discovered and read. The discovered iBeacons' Major and Minor numbers are looked up to learn that the user is in a specific room on a specific floor in the museum. Relevant content is accessed from the museum's API and loaded in the smartphone app.

Example

An Android app acting as a Central locates its position relative to 3 or more nearby iBeacons in a room (Figure 9-5).

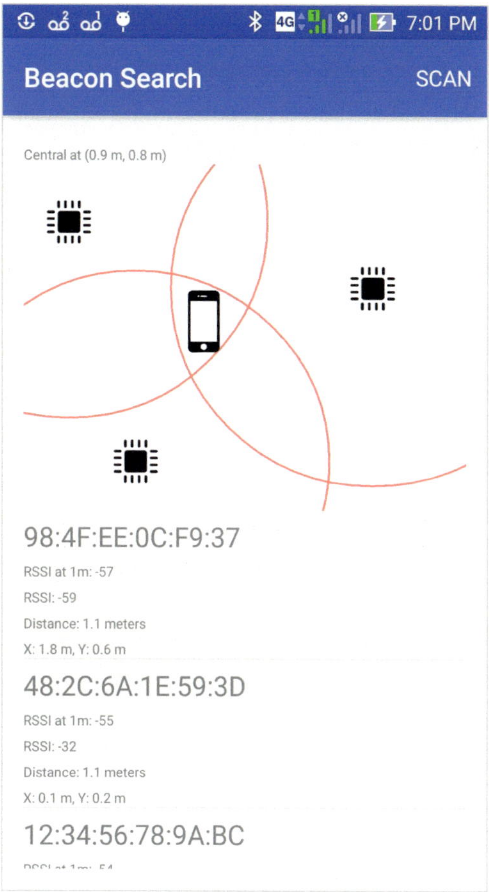

Figure 9-5. Android App locates its own position relative to 3 or more nearby iBeacons in a room.

Example: Echo Client and Server

An Echo Server is the “Hello World” of network programming. It has the minimum features required to transmit, store, and respond to data on a network - the core features required for any network application.

And yet it must support all the features you’ve learned so far in this book - advertising, reads, writes, notifications, segmented data transfer, and encryption. It’s a sophisticated program!

The Echo Server works like this:

In this example, the Peripheral acts as a server, the “Echo Server” and the Central acts as a client (Figure 10-1).

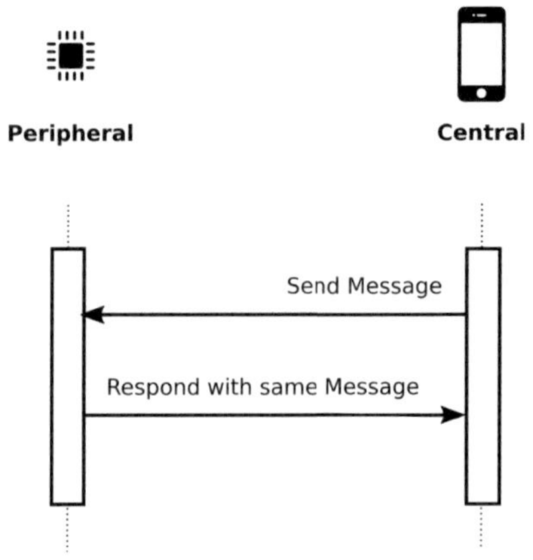

Figure 10-1. How an Echo Server works

Examples

A smartphone can act as a Central in this example. An iPhone app acting as a Central that shows a message queued to send to an Echo Server (Figure 9-2), and that message echoed back after being sent (Figure 9-3).

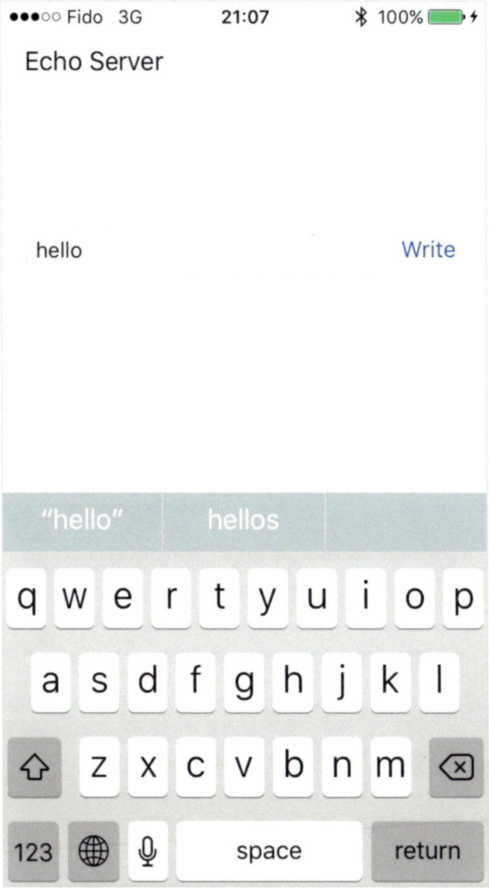

Figure 9-2. Message queued to be sent to a Bluetooth Low Energy Echo Server

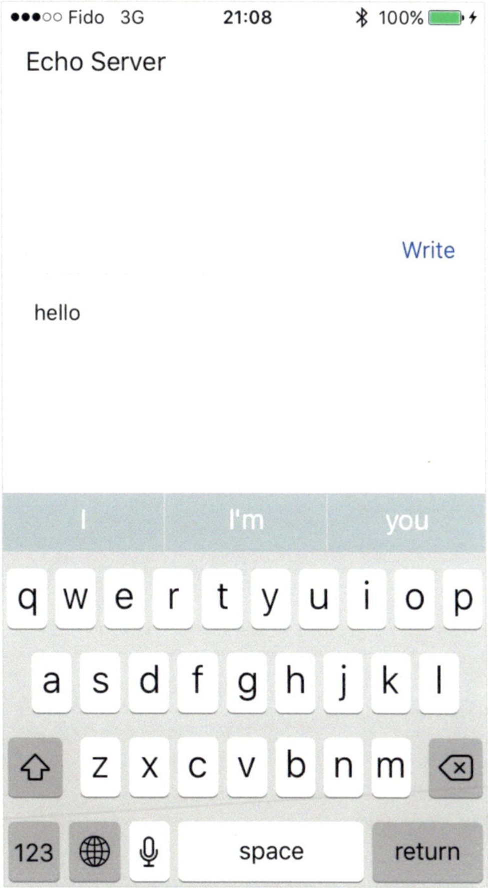

Figure 9-3. Message echoed back from the Bluetooth Echo Server

The Peripheral in this example may process the inbound text behind the scenes and echo it back. If the Peripheral were connected to some kind of debugging console, it may report something like this (Figure 10-4).

```
Starting EchoServer
Incoming message found: hello
Sending message: hello
```

Figure 10-4. Echo Server debugging console output

Example: Remote Control LED

So far, this book has worked a lot with text data rather than binary data, because it's easy to text without using specialized tools such as oscilloscopes or logic analyzers.

Most real-world projects transmit binary instead of text. Binary is much more efficient in transmitting information.

Because binary data it is the language of computers, it is easier to work with than text. There is no need to worry about character sets, null characters, or cut-off words.

This project will show how to remotely control an LED on a Peripheral using software on a Central.

The LED Remote works like this (Figure 11-1).

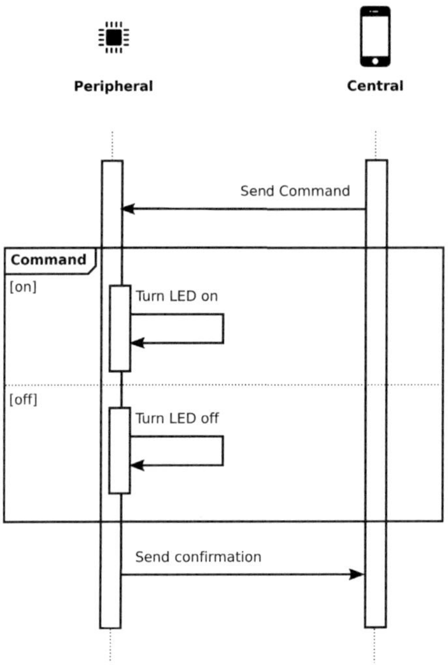

Figure 10-1. How a Remote Control LED works

In all the other examples, text was being sent between Central and Peripheral.

In order for the Central and Peripheral to understand each other, they need shared language between them. In this case, a data packet format.

Sending Commands to Peripheral

When the Central sends a message, it should be able to specify if it is sending a command or an error. We can do this in two bytes, like this (Figure 11-2).

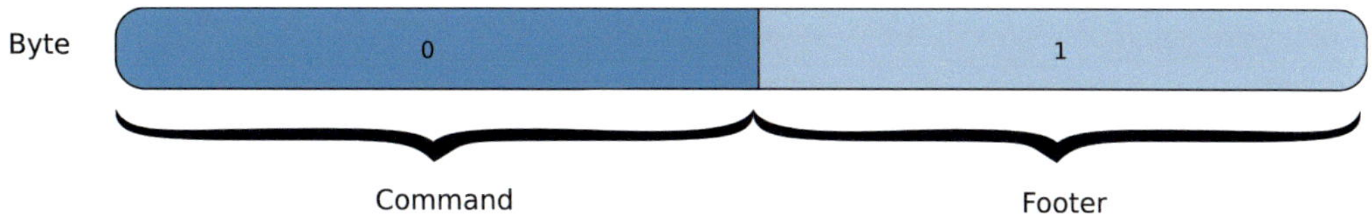

Figure 11-2. Packet structure for commands

The Peripheral reads the footer byte of the incoming message to determine the type of message, i.e., an error or a command. For example, define the message types as:

Table 11-1. Footer Values

Name	Value	Description
bleResponseError	0	The Central is sending an error
bleResponseConfirmation	1	The Central is sending a confirmation
bleResponseCommand	2	The Central is sending a command

The Peripheral reads the first byte to determine the type of error or command. For example, define the commands as:

Table 11-2. Command Values

Name	Value	Description
bleCommandLedOff	1	Turn off the Peripheral's LED
bleCommandLedOn	2	Turn on the Peripheral's LED

The Peripheral then responds to the Central with a status message regarding the success or failure to execute the command. This can also be expressed as two bytes (Figure 11-3).

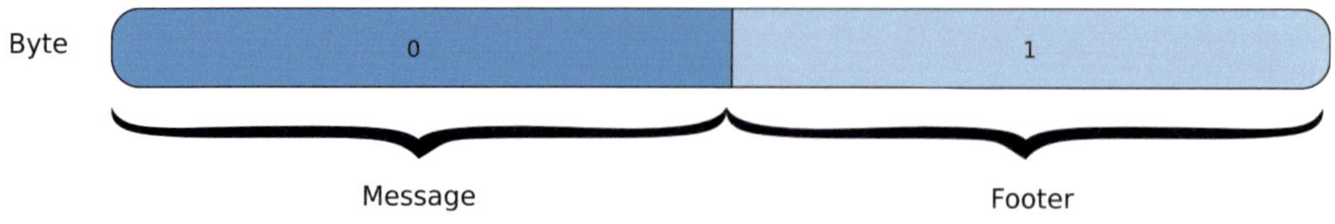

Figure 11-3. Packet structure for responses

If the Peripheral sends a confirmation that the LED state has changed, then the Central inspects the first byte of the message to determine what the current state of the Peripheral's LED is:

Table 11-2. Confirmation Values

Name	Value	Description
ledStateOff	1	The Peripheral's LED is off
ledStateOn	2	The Peripheral's LED is on

In this way, a common language is established between the Central and the Peripheral.

Gatt Profile

The Bluetooth Low Energy specification provides a special Service, the Automation IO Service (0x1815), specifically for remote control devices such as this.

It is a best practice to use each Characteristic for a single purpose. For this reason, Characteristic 0x2a56 will be used for sending commands to the Peripheral and Characteristic 0x2a57 will be used for responses from the Peripheral:

Table 11-4. Characteristic Usages

UUID	Use
0x2a56	Send commands from Central to Peripheral
0x2a57	Send responses from Peripheral to Central

Example

An iPhone app acting as a Central connected to a remote controllable LED with the LED in the off state (Figure 11-4) and in the on state (Figure 11-5).

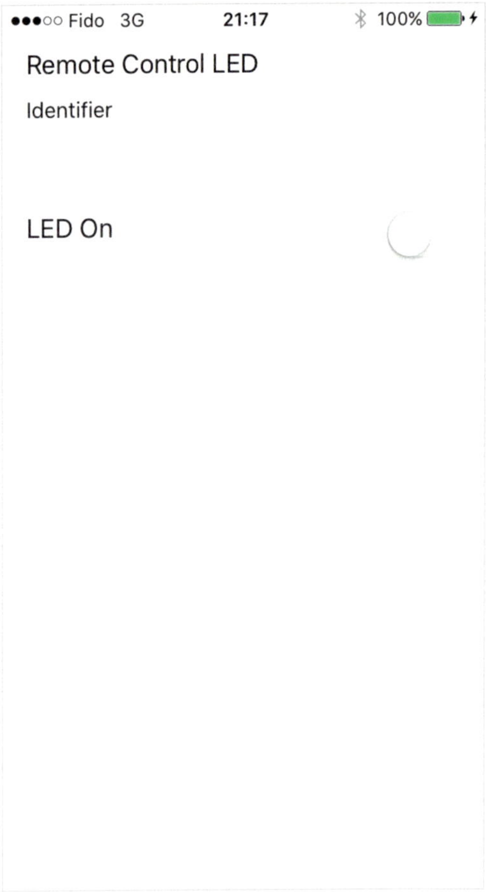

Figure 11-2. Remote LED in "off" state

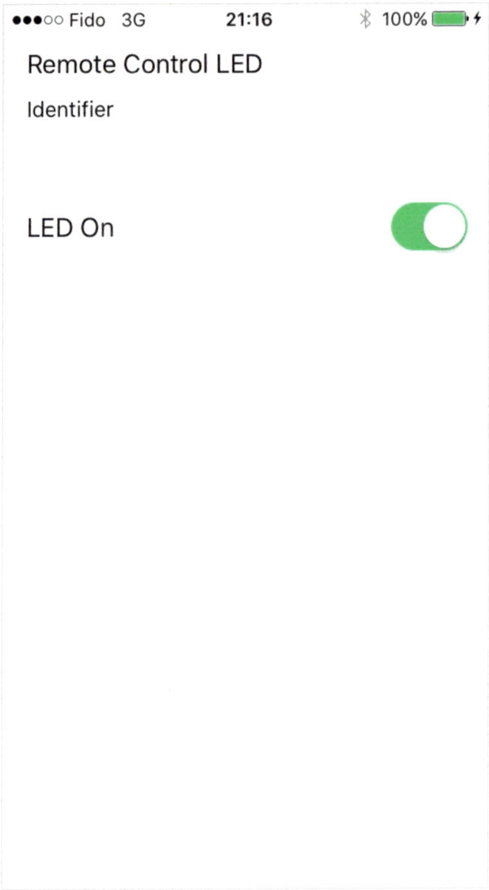

Figure 11-3. Remote LED in "on" state

On the Peripheral side, one example is to have an Arduino prototyping board turn its onboard LED on when it receives the command to do so from a connected Central (Figure 11-4).

Figure 11-4 a prototyping board shining the onboard LED in response to a programmed Bluetooth command to turn it on

Appendix

For reference, the following are properties of the Bluetooth Low Energy network and hardware.

Range	100 m (330 ft)
Data Rate	1M bit/s
Application Throughput	0.27 Mbit/s
Security	128-bit AES with Counter Mode CBC-MAC and application layer user defined (BEWARE: this encryption has vulnerabilities)
Robustness	Adaptive Frequency Hopping, Lazy Acknowledgement, 24-bit CRC, 32-bit Message Integrity Check
Range	100 m (330 ft)
Data Rate	1M bit/s
Application Throughput	0.27 Mbit/s
Security	128-bit AES with Counter Mode CBC-MAC and application layer user defined (BEWARE: this encryption has vulnerabilities)
Peak Current Consumption	< 15 mA
Byte-Order in Broadcast	Big Endian (most significant bit at end)
Range	100 m (330 ft)
Data Rate	1M bit/s
Application Throughput	0.27 Mbit/s
Security	128-bit AES with Counter Mode CBC-MAC and application layer user defined (BEWARE: this encryption has vulnerabilities)

Appendix II: UUID Format

Bluetooth Low Energy has tight space requirements. Therefore it is preferred to transmit 16-bit UUIDs instead of 32-bit UUIDs. UUIDs can be converted between 16-bit and 32-bit with the standard Bluetooth Low Energy UUID format:

Table II-1. 16-bit to 32-bit UUID Conversion Standard

UUID Format	uuid16	Resulting uuid32
0000*0000*-0000-1000-8000-00805f9b34fb	0x2A56	0000*2A56*-0000-1000-8000-00805f9b34fb

Appendix III: Minimal Recommended GATT

As a best practice, it is good to host a standard set of Services and Characteristics in a Peripheral's GATT Profile. These Characteristics allow connected Centrals to get the make and model number of the device, and the battery level if the Peripheral is battery-powered:

Table III-1. Minimal GATT Profile

GATT Type	Name	Data Type	UUID
Service	Device Information Service		0x180a
Characteristic	Device Name	char array	0x2a00
Characteristic	Model Number	char array	0x2a24
Characteristic	Serial Number	char array	0x2a04
Service	Battery Level		0x180f
Characteristic	Battery Level	integer	0x2a19

Appendix IV: Reserved GATT Services

Services act as a container for Characteristics or other Services, providing a tree-like structure for organizing Bluetooth I/O.

These Services UUIDs have been reserved for special contexts, such as Device Information (0x180A) Which may contain Characteristics that communicate information about the Peripheral's name, version number, or settings.

Note: All Bluetooth Peripherals should have a Battery Service (0x180F) Service containing a Battery Level (0x2A19) Characteristic.

Table IV-1. Reserved GATT Services

Specification Name	UUID	Specification Type
Alert Notification Service	0x1811	org.bluetooth.service.alert_notification
Automation IO	0x1815	org.bluetooth.service.automation_io
Battery Service	0x180F	org.bluetooth.service.battery_service
Blood Pressure	0x1810	org.bluetooth.service.blood_pressure
Body Composition	0x181B	org.bluetooth.service.body_composition
Bond Management	0x181E	org.bluetooth.service.bond_management
Continuous Glucose Monitoring	0x181F	org.bluetooth.service.continuous_glucose_monitoring

Specification Name	UUID	Specification Type
Current Time Service	0x1805	org.bluetooth.service.current_time
Cycling Power	0x1818	org.bluetooth.service.cycling_power
Cycling Speed and Cadence	0x1816	org.bluetooth.service.cycling_speed_and_cadence
Device Information	0x180A	org.bluetooth.service.device_information
Environmental Sensing	0x181A	org.bluetooth.service.environmental_sensing
Generic Access	0x1800	org.bluetooth.service.generic_access
Generic Attribute	0x1801	org.bluetooth.service.generic_attribute
Glucose	0x1808	org.bluetooth.service.glucose
Health Thermometer	0x1809	org.bluetooth.service.health_thermometer
Heart Rate	0x180D	org.bluetooth.service.heart_rate
HTTP Proxy	0x1823	org.bluetooth.service.http_proxy
Human Interface Device	0x1812	org.bluetooth.service.human_interface_device
Immediate Alert	0x1802	org.bluetooth.service.immediate_alert
Indoor Positioning	0x1821	org.bluetooth.service.indoor_positioning
Internet Protocol Support	0x1820	org.bluetooth.service.internet_protocol_support

Specification Name	UUID	Specification Type
Link Loss	0x1803	org.bluetooth.service.link_loss
Location and Navigation	0x1819	org.bluetooth.service.location_and_navigation
Next DST Change Service	0x1807	org.bluetooth.service.next_dst_change
Object Transfer	0x1825	org.bluetooth.service.object_transfer
Phone Alert Status Service	0x180E	org.bluetooth.service.phone_alert_status
Pulse Oximeter	0x1822	org.bluetooth.service.pulse_oximeter
Reference Time Update Service	0x1806	org.bluetooth.service.reference_time_update
Running Speed and Cadence	0x1814	org.bluetooth.service.running_speed_and_cadence
Scan Parameters	0x1813	org.bluetooth.service.scan_parameters
Transport Discovery	0x1824	org.bluetooth.service.transport_discovery
Tx Power	0x1804	org.bluetooth.service.tx_power
User Data	0x181C	org.bluetooth.service.user_data
Weight Scale	0x181D	org.bluetooth.service.weight_scale

Source: Bluetooth SIG: GATT Services
Retrieved from https://www.bluetooth.com/specifications/gatt/services

Appendix V: Reserved GATT Characteristics

Characteristics act a data port that can be read from or written to.

These Characteristic UUIDs have been reserved for specific types of data, such as Device Name (0x2A00) which may read the Peripheral's current battery level.

Note: All Bluetooth Peripherals should have a Battery Level (0x2A19) Characteristic, contained inside a Battery Service (0x180F) Service.

Table V-1. Reserved GATT Characteristics

Specification Name	UUID	Specification Type
Aerobic Heart Rate Lower Limit	0x2A7E	org.bluetooth.characteristic.aerobic_heart_rate_lower_limit
Aerobic Heart Rate Upper Limit	0x2A84	org.bluetooth.characteristic.aerobic_heart_rate_upper_limit
Aerobic Threshold	0x2A7F	org.bluetooth.characteristic.aerobic_threshold
Age	0x2A80	org.bluetooth.characteristic.age
Aggregate	0x2A5A	org.bluetooth.characteristic.aggregate
Alert Category ID	0x2A43	org.bluetooth.characteristic.alert_category_id
Alert Category ID Bit Mask	0x2A42	org.bluetooth.characteristic.alert_category_id_bit_mask
Alert Level	0x2A06	org.bluetooth.characteristic.alert_level

Specification Name	UUID	Specification Type
Alert Status	0x2A3F	org.bluetooth.characteristic.alert_status
Altitude	0x2AB3	org.bluetooth.characteristic.altitude
Anaerobic Heart Rate Lower Limit	0x2A81	org.bluetooth.characteristic.anaerobic_heart_rate_lower_limit
Anaerobic Heart Rate Upper Limit	0x2A82	org.bluetooth.characteristic.anaerobic_heart_rate_upper_limit
Anaerobic Threshold	0x2A83	org.bluetooth.characteristic.anaerobic_threshold
Analog	0x2A58	org.bluetooth.characteristic.analog
Apparent Wind Direction	0x2A73	org.bluetooth.characteristic.apparent_wind_direction
Apparent Wind Speed	0x2A72	org.bluetooth.characteristic.apparent_wind_speed
Appearance	0x2A01	org.bluetooth.characteristic.gap.appearance
Barometric Pressure Trend	0x2AA3	org.bluetooth.characteristic.barometric_pressure_trend
Battery Level	0x2A19	org.bluetooth.characteristic.battery_level
Blood Pressure Feature	0x2A49	org.bluetooth.characteristic.blood_pressure_feature
Blood Pressure Measurement	0x2A35	org.bluetooth.characteristic.blood_pressure_measurement
Body Composition Feature	0x2A9B	org.bluetooth.characteristic.body_composition_feature
Body Composition Measurement	0x2A9C	org.bluetooth.characteristic.body_composition_measurement
Body Sensor Location	0x2A38	org.bluetooth.characteristic.body_sensor_location

Specification Name	UUID	Specification Type
Bond Management Control Point	0x2AA4	org.bluetooth.characteristic.bond_management_control_point
Bond Management Feature	0x2AA5	org.bluetooth.characteristic.bond_management_feature
Boot Keyboard Input Report	0x2A22	org.bluetooth.characteristic.boot_keyboard_input_report
Boot Keyboard Output Report	0x2A32	org.bluetooth.characteristic.boot_keyboard_output_report
Boot Mouse Input Report	0x2A33	org.bluetooth.characteristic.boot_mouse_input_report
Central Address Resolution	0x2AA6	org.bluetooth.characteristic.gap.central_address_resolution_support
CGM Feature	0x2AA8	org.bluetooth.characteristic.cgm_feature
CGM Measurement	0x2AA7	org.bluetooth.characteristic.cgm_measurement
CGM Session Run Time	0x2AAB	org.bluetooth.characteristic.cgm_session_run_time
CGM Session Start Time	0x2AAA	org.bluetooth.characteristic.cgm_session_start_time
CGM Specific Ops Control Point	0x2AAC	org.bluetooth.characteristic.cgm_specific_ops_control_point
CGM Status	0x2AA9	org.bluetooth.characteristic.cgm_status
CSC Feature	0x2A5C	org.bluetooth.characteristic.csc_feature
CSC Measurement	0x2A5B	org.bluetooth.characteristic.csc_measurement
Current Time	0x2A2B	org.bluetooth.characteristic.current_time

Specification Name	UUID	Specification Type
Cycling Power Feature	0x2A65	org.bluetooth.characteristic.cycling_power_feature
Cycling Power Measurement	0x2A63	org.bluetooth.characteristic.cycling_power_measurement
Cycling Power Vector	0x2A64	org.bluetooth.characteristic.cycling_power_vector
Database Change Increment	0x2A99	org.bluetooth.characteristic.database_change_increment
Date of Birth	0x2A85	org.bluetooth.characteristic.date_of_birth
Date of Threshold Assessment	0x2A86	org.bluetooth.characteristic.date_of_threshold_assessment
Date Time	0x2A08	org.bluetooth.characteristic.date_time
Day Date Time	0x2A0A	org.bluetooth.characteristic.day_date_time
Day of Week	0x2A09	org.bluetooth.characteristic.day_of_week
Descriptor Value Changed	0x2A7D	org.bluetooth.characteristic.descriptor_value_changed
Device Name	0x2A00	org.bluetooth.characteristic.gap.device_name
Dew Point	0x2A7B	org.bluetooth.characteristic.dew_point
Digital	0x2A56	org.bluetooth.characteristic.digital
DST Offset	0x2A0D	org.bluetooth.characteristic.dst_offset
Elevation	0x2A6C	org.bluetooth.characteristic.elevation

Specification Name	UUID	Specification Type
Exact Time 256	0x2A0C	org.bluetooth.characteristic.exact_time_256
Fat Burn Heart Rate Lower Limit	0x2A88	org.bluetooth.characteristic.fat_burn_heart_rate_lower_limit
Fat Burn Heart Rate Upper Limit	0x2A89	org.bluetooth.characteristic.fat_burn_heart_rate_upper_limit
Firmware Revision String	0x2A26	org.bluetooth.characteristic.firmware_revision_string
First Name	0x2A8A	org.bluetooth.characteristic.first_name
Five Zone Heart Rate Limits	0x2A8B	org.bluetooth.characteristic.five_zone_heart_rate_limits
Floor Number	0x2AB2	org.bluetooth.characteristic.floor_number
Gender	0x2A8C	org.bluetooth.characteristic.gender
Glucose Feature	0x2A51	org.bluetooth.characteristic.glucose_feature
Glucose Measurement	0x2A18	org.bluetooth.characteristic.glucose_measurement
Glucose Measurement Context	0x2A34	org.bluetooth.characteristic.glucose_measurement_context
Gust Factor	0x2A74	org.bluetooth.characteristic.gust_factor
Hardware Revision String	0x2A27	org.bluetooth.characteristic.hardware_revision_string
Heart Rate Control Point	0x2A39	org.bluetooth.characteristic.heart_rate_control_point
Heart Rate Max	0x2A8D	org.bluetooth.characteristic.heart_rate_max
Heart Rate Measurement	0x2A37	org.bluetooth.characteristic.heart_rate_measurement

Specification Name	UUID	Specification Type
Heat Index	0x2A7A	org.bluetooth.characteristic.heat_index
Height	0x2A8E	org.bluetooth.characteristic.height
HID Control Point	0x2A4C	org.bluetooth.characteristic.hid_control_point
HID Information	0x2A4A	org.bluetooth.characteristic.hid_information
Hip Circumference	0x2A8F	org.bluetooth.characteristic.hip_circumference
HTTP Control Point	0x2ABA	org.bluetooth.characteristic.http_control_point
HTTP Entity Body	0x2AB9	org.bluetooth.characteristic.http_entity_body
HTTP Headers	0x2AB7	org.bluetooth.characteristic.http_headers
HTTP Status Code	0x2AB8	org.bluetooth.characteristic.http_status_code
HTTPS Security	0x2ABB	org.bluetooth.characteristic.https_security
Humidity	0x2A6F	org.bluetooth.characteristic.humidity
IEEE 11073-20601 Regulatory Certification Data List	0x2A2A	org.bluetooth.characteristic.ieee_11073-20601_regulatory_certification_data_list
Indoor Positioning Configuration	0x2AAD	org.bluetooth.characteristic.indoor_positioning_configuration
Intermediate Cuff Pressure	0x2A36	org.bluetooth.characteristic.intermediate_cuff_pressure
Intermediate Temperature	0x2A1E	org.bluetooth.characteristic.intermediate_temperature

Specification Name	UUID	Specification Type
Language	0x2AA2	org.bluetooth.characteristic.language
Last Name	0x2A90	org.bluetooth.characteristic.last_name
Latitude	0x2AAE	org.bluetooth.characteristic.latitude
LN Control Point	0x2A6B	org.bluetooth.characteristic.ln_control_point
LN Feature	0x2A6A	org.bluetooth.characteristic.ln_feature
Local East Coordinate	0x2AB1	org.bluetooth.characteristic.local_east_coordinate
Local North Coordinate	0x2AB0	org.bluetooth.characteristic.local_north_coordinate
Local Time Information	0x2A0F	org.bluetooth.characteristic.local_time_information
Location and Speed	0x2A67	org.bluetooth.characteristic.location_and_speed
Location Name	0x2AB5	org.bluetooth.characteristic.location_name
Longitude	0x2AAF	org.bluetooth.characteristic.longitude
Magnetic Declination	0x2A2C	org.bluetooth.characteristic.magnetic_declination
Magnetic Flux Density - 2D	0x2AA0	org.bluetooth.characteristic.magnetic_flux_density_2D
Magnetic Flux Density - 3D	0x2AA1	org.bluetooth.characteristic.magnetic_flux_density_3D
Manufacturer Name String	0x2A29	org.bluetooth.characteristic.manufacturer_name_string
Maximum Recommended Heart Rate	0x2A91	org.bluetooth.characteristic.maximum_recommended_heart_rate

Specification Name	UUID	Specification Type
Measurement Interval	0x2A21	org.bluetooth.characteristic.measurement_interval
Model Number String	0x2A24	org.bluetooth.characteristic.model_number_string
Navigation	0x2A68	org.bluetooth.characteristic.navigation
New Alert	0x2A46	org.bluetooth.characteristic.new_alert
Object Action Control Point	0x2AC5	org.bluetooth.characteristic.object_action_control_point
Object Changed	0x2AC8	org.bluetooth.characteristic.object_changed
Object First-Created	0x2AC1	org.bluetooth.characteristic.object_first_created
Object ID	0x2AC3	org.bluetooth.characteristic.object_id
Object Last-Modified	0x2AC2	org.bluetooth.characteristic.object_last_modified
Object List Control Point	0x2AC6	org.bluetooth.characteristic.object_list_control_point
Object List Filter	0x2AC7	org.bluetooth.characteristic.object_list_filter
Object Name	0x2ABE	org.bluetooth.characteristic.object_name
Object Properties	0x2AC4	org.bluetooth.characteristic.object_properties
Object Size	0x2AC0	org.bluetooth.characteristic.object_size
Object Type	0x2ABF	org.bluetooth.characteristic.object_type
OTS Feature	0x2ABD	org.bluetooth.characteristic.ots_feature

Specification Name	UUID	Specification Type
Peripheral Preferred Connection Parameters	0x2A04	org.bluetooth.characteristic.gap.peripheral_preferred_connection_parameters
Peripheral Privacy Flag	0x2A02	org.bluetooth.characteristic.gap.peripheral_privacy_flag
PLX Continuous Measurement	0x2A5F	org.bluetooth.characteristic.plx_continuous_measurement
PLX Features	0x2A60	org.bluetooth.characteristic.plx_features
PLX Spot-Check Measurement	0x2A5E	org.bluetooth.characteristic.plx_spot_check_measurement
PnP ID	0x2A50	org.bluetooth.characteristic.pnp_id
Pollen Concentration	0x2A75	org.bluetooth.characteristic.pollen_concentration
Position Quality	0x2A69	org.bluetooth.characteristic.position_quality
Pressure	0x2A6D	org.bluetooth.characteristic.pressure
Protocol Mode	0x2A4E	org.bluetooth.characteristic.protocol_mode
Rainfall	0x2A78	org.bluetooth.characteristic.rainfall
Reconnection Address	0x2A03	org.bluetooth.characteristic.gap.reconnection_address
Record Access Control Point	0x2A52	org.bluetooth.characteristic.record_access_control_point
Reference Time Information	0x2A14	org.bluetooth.characteristic.reference_time_information
Report	0x2A4D	org.bluetooth.characteristic.report

Specification Name	UUID	Specification Type
Resolvable Private Address Only	0x2AC9	org.bluetooth.characteristic.resolvable_private_address_only
Resting Heart Rate	0x2A92	org.bluetooth.characteristic.resting_heart_rate
Ringer Control Point	0x2A40	org.bluetooth.characteristic.ringer_control_point
Ringer Setting	0x2A41	org.bluetooth.characteristic.ringer_setting
RSC Feature	0x2A54	org.bluetooth.characteristic.rsc_feature
RSC Measurement	0x2A53	org.bluetooth.characteristic.rsc_measurement
SC Control Point	0x2A55	org.bluetooth.characteristic.sc_control_point
Scan Interval Window	0x2A4F	org.bluetooth.characteristic.scan_interval_window
Scan Refresh	0x2A31	org.bluetooth.characteristic.scan_refresh
Sensor Location	0x2A5D	org.blueooth.characteristic.sensor_location
Serial Number String	0x2A25	org.bluetooth.characteristic.serial_number_string
Service Changed	0x2A05	org.bluetooth.characteristic.gatt.service_changed
Software Revision String	0x2A28	org.bluetooth.characteristic.software_revision_string
Sport Type for Aerobic and Anaerobic Thresholds	0x2A93	org.bluetooth.characteristic.sport_type_for_aerobic_and_anaerobic_t hresholds
Supported New Alert Category	0x2A47	org.bluetooth.characteristic.supported_new_alert_category

Specification Name	UUID	Specification Type
System ID	0x2A23	org.bluetooth.characteristic.system_id
TDS Control Point	0x2ABC	org.bluetooth.characteristic.tds_control_point
Temperature	0x2A6E	org.bluetooth.characteristic.temperature
Temperature Measurement	0x2A1C	org.bluetooth.characteristic.temperature_measurement
Temperature Type	0x2A1D	org.bluetooth.characteristic.temperature_type
Three Zone Heart Rate Limits	0x2A94	org.bluetooth.characteristic.three_zone_heart_rate_limits
Time Accuracy	0x2A12	org.bluetooth.characteristic.time_accuracy
Time Source	0x2A13	org.bluetooth.characteristic.time_source
Time Update Control Point	0x2A16	org.bluetooth.characteristic.time_update_control_point
Time Update State	0x2A17	org.bluetooth.characteristic.time_update_state
Time with DST	0x2A11	org.bluetooth.characteristic.time_with_dst
Time Zone	0x2A0E	org.bluetooth.characteristic.time_zone
True Wind Direction	0x2A71	org.bluetooth.characteristic.true_wind_direction
True Wind Speed	0x2A70	org.bluetooth.characteristic.true_wind_speed
Two Zone Heart Rate Limit	0x2A95	org.bluetooth.characteristic.two_zone_heart_rate_limit
Tx Power Level	0x2A07	org.bluetooth.characteristic.tx_power_level

Specification Name	UUID	Specification Type
Uncertainty	0x2AB4	org.bluetooth.characteristic.uncertainty
Unread Alert Status	0x2A45	org.bluetooth.characteristic.unread_alert_status
URI	0x2AB6	org.bluetooth.characteristic.uri
User Control Point	0x2A9F	org.bluetooth.characteristic.user_control_point
User Index	0x2A9A	org.bluetooth.characteristic.user_index
UV Index	0x2A76	org.bluetooth.characteristic.uv_index
VO2 Max	0x2A96	org.bluetooth.characteristic.vo2_max
Waist Circumference	0x2A97	org.bluetooth.characteristic.waist_circumference
Weight	0x2A98	org.bluetooth.characteristic.weight
Weight Measurement	0x2A9D	org.bluetooth.characteristic.weight_measurement
Weight Scale Feature	0x2A9E	org.bluetooth.characteristic.weight_scale_feature
Wind Chill	0x2A79	org.bluetooth.characteristic.wind_chill

Source: Bluetooth SIG: GATT Characteristics
Retrieved from https://www.bluetooth.com/specifications/gatt/characteristics

Appendix VI: GATT Descriptors

The following GATT Descriptor UUIDs have been reserved for specific uses.

GATT Descriptors describe features within a Characteristic that can be altered, for instance, the Client Characteristic Configuration (0x2902) which can be flagged to allow a connected Central to subscribe to notifications on a Characteristic.

Table VI-1. Reserved GATT Descriptors

Specification Name	UUID	Specification Type
Characteristic Aggregate Format	0x2905	org.bluetooth.descriptor.gatt.characteristic_aggregate_format
Characteristic Extended Properties	0x2900	org.bluetooth.descriptor.gatt.characteristic_extended_properties
Characteristic Presentation Format	0x2904	org.bluetooth.descriptor.gatt.characteristic_presentation_format
Characteristic User Description	0x2901	org.bluetooth.descriptor.gatt.characteristic_user_description
Client Characteristic Configuration	0x2902	org.bluetooth.descriptor.gatt.client_characteristic_configuration
Environmental Sensing Configuration	0x290B	org.bluetooth.descriptor.es_configuration
Environmental Sensing Measurement	0x290C	org.bluetooth.descriptor.es_measurement

Specification Name	UUID	Specification Type
Number of Digitals	0x2909	org.bluetooth.descriptor.number_of_digitals
Report Reference	0x2908	org.bluetooth.descriptor.report_reference
Server Characteristic Configuration	0x2903	org.bluetooth.descriptor.gatt.server_characteristic_configuration
Time Trigger Setting	0x290E	org.bluetooth.descriptor.time_trigger_setting
Valid Range	0x2906	org.bluetooth.descriptor.valid_range
Value Trigger Setting	0x290A	org.bluetooth.descriptor.value_trigger_setting

Source: Bluetooth SIG: GATT Descriptors

Retrieved from https://www.bluetooth.com/specifications/gatt/descriptors

Appendix VII: Company Identifiers

The following companies have specific Manufacturer Identifiers, which identify Bluetooth devices in the Generic Access Profile (GAP). Peripherals with no specific manufacturer use ID 65535 (0xffff). All other IDs are reserved, even if not yet assigned.

This is a non-exhaustive list of companies. A full list and updated can be found on the Bluetooth SIG website.

Table VII-1. Company Identifiers

Decimal	Hexadecimal	Company
0	0x0000	Ericsson Technology Licensing
1	0x0001	Nokia Mobile Phones
2	0x0002	Intel Corp.
3	0x0003	IBM Corp.
4	0x0004	Toshiba Corp.
5	0x0005	3Com
6	0x0006	Microsoft
7	0x0007	Lucent

Decimal	Hexadecimal	Company
8	0x0008	Motorola
13	0x000D	Texas Instruments Inc.
19	0x0013	Atmel Corporation
29	0x001D	Qualcomm
36	0x0024	Alcatel
37	0x0025	NXP Semiconductors (formerly Philips Semiconductors)
60	0x003C	BlackBerry Limited (formerly Research In Motion)
76	0x004C	Apple, Inc.
86	0x0056	Sony Ericsson Mobile Communications
89	0x0059	Nordic Semiconductor ASA
92	0x005C	Belkin International, Inc.
93	0x005D	Realtek Semiconductor Corporation
101	0x0065	Hewlett-Packard Company
104	0x0068	General Motors
117	0x0075	Samsung Electronics Co. Ltd.

Decimal	Hexadecimal	Company
120	0x0078	Nike, Inc.
135	0x0087	Garmin International, Inc.
138	0x008A	Jawbone
184	0x00B8	Qualcomm Innovation Center, Inc. (QuIC)
215	0x00D7	Qualcomm Technologies, Inc.
216	0x00D8	Qualcomm Connected Experiences, Inc.
220	0x00DC	Procter & Gamble
224	0x00E0	Google
359	0x0167	Bayer HealthCare
367	0x016F	Podo Labs, Inc
369	0x0171	Amazon Fulfillment Service
387	0x0183	Walt Disney
398	0x018E	Fitbit, Inc.
425	0x01A9	Canon Inc.
427	0x01AB	Facebook, Inc.

Decimal	Hexadecimal	Company
474	0x01DA	Logitech International SA
558	0x022E	Siemens AG
605	0x025D	Lexmark International Inc.
637	0x027D	HUAWEI Technologies Co., Ltd. ( )
720	0x02D0	3M
876	0x036C	Zipcar
897	0x0381	Sharp Corporation
921	0x0399	Nikon Corporation
1117	0x045D	Boston Scientific Corporation
65535	0xFFFF	No Device ID

Source: Bluetooth SIG: Company Identifiers
Retrieved from
https://www.bluetooth.com/specifications/assigned-numbers/company-identifiers

Glossary

The following is a list of Bluetooth Low Energy terms and their meanings.

Android - An open-source operating system used for smartphones and tablet computers

Arduino - An open source computer hardware and software company, project, and user community that designs and manufactures single-board microcontrollers and microcontroller kits for building digital devices and interactive objects that can sense and control objects in the physical world.

Attribute - An unit of a GATT Profile which can be accessed by a Central, such as a Service or a Characteristic.

Beacon - A Bluetooth Low Energy Peripheral which continually Broadcasts so that Centrals can discern their location from information gleaned from the properties of the broadcast.

Bluetooth Low Energy (BLE) - A low power, short range wireless protocol used on micro electronics.

Broadcast - A feature of Bluetooth Low Energy where a Peripheral outputs a name and other specific data about a itself

Central - A Bluetooth Low Energy device that can connect to several Peripherals.

Channel - A finely-tuned radio frequency used for Broadcasting or data transmission.

Characteristic - A port or data endpoint where data can be read or written.

Descriptor - A feature of a Characteristic that allows for some sort of data interaction, such as Read, Write, or Notify.

E0 - The encryption algorithm built into Bluetooth Low Energy.

Generic Attribute (GATT) Profile - A list of Services and Characteristics which are unique to a Peripheral and describe how data is served from the Peripheral. GATT profiles are hosted by a Peripheral

Intel® Curie™ Module - The Intel® module that powers the Arduino 101 and contains the Bluetooth chipset.

iBeacon - An Apple compatible Beacon which allows a Central to download a specific packet of data to inform the Central of its absolute location and other properties.

Notify - An operation where a Peripheral alerts a Central of a change in data.

nRFx - A series of Bluetooth-enabled programmable microcontrollers produced by Nordic Semiconductors®.

Peripheral - A Bluetooth Low Energy device that can connect to a single Central. Peripherals host a Generic Attribute (GATT) profile.

Read - An operation where a Central downloads data from a Characteristic.

Scan - The process of a Central searching for Broadcasting Peripherals.

Scan Response - A feature of Bluetooth Low Energy which allows Centrals to download a small packet of data without connecting.

Service - A container structure used to organize data endpoints. Services are hosted by a Peripheral.

Universally Unique Identifier (UUID) - A long, randomly generated alphanumeric sting that is unique regardless of where it's used. UUIDs are designed to avoid name collisions that may happen when countless programs are interacting with each other.

Write - An operation where a Central alters data on a Characteristic.

(This page intentionally left blank)

About the Author

Tony's infinite curiosity compels him to want to open up and learn about everything he touches, and his excitement compels him to share what he learns with others.

He has two true passions: branding and inventing.

His passion for branding led him to start a company that did branding and marketing in 4 countries for firms such as Apple, Intel, and Sony BMG. He loves weaving the elements of design, writing, product, and strategy into an essential truth that defines a company.

His passion for inventing led him to start a company that uses brain imaging to quantify meditation and to predict seizures, a company acquired $1.5m in funding and was incubated in San Francisco where he currently resides.

Those same passions have led him on some adventures as well, including living in a Greek monastery with orthodox monks and to tagging along with a gypsy in Spain to learn to play flamenco guitar.

(This page intentionally left blank)

About this Book

In this book, you will learn the mechanics behind the Bluetooth Low Energy protocol, a wireless technology that lets computers, smartphones, and smart devices communicate within a space about the size of a room.

Through the course of the book you will learn important concepts that relate to:

- How Bluetooth Low Energy works,
- How data is sent and received
- Common paradigms for handling data

Additionally, you will learn how three common product paradigms work:

- An iBeacon
- An Echo Server
- A Remote Controlled Device

Skill Level

This book is an excellent read for anyone who wants to know the terminology, technology, and concepts behind Bluetooth Low Energy devices but doesn't need to program them.

Other Books in this Series

If you are interested in programming Bluetooth Low Energy Devices, please check out the other books in this series or visit bluetoothlowenergybooks.com:

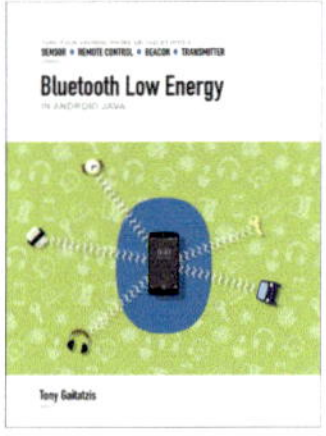

Bluetooth Low Energy in Android Java

Tony Gaitatzis, 2017

ISBN: 978-1-7751280-4-5

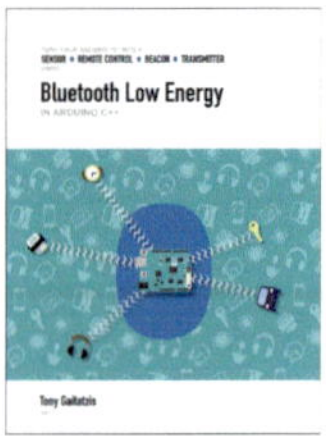

Bluetooth Low Energy in Arduino 101

Tony Gaitatzis, 2017

ISBN: 978-1-7751280-6-9

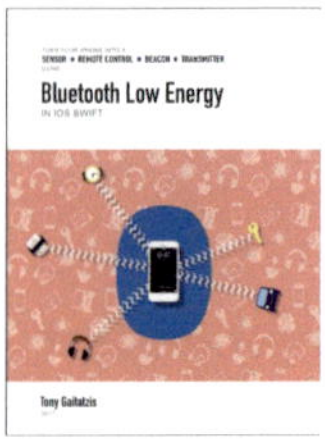

Bluetooth Low Energy in iOS Swift

Tony Gaitatzis, 2017

ISBN: 978-1-7751280-5-2

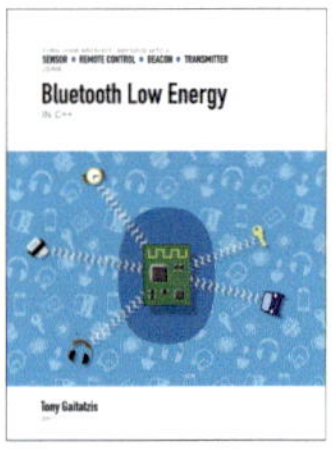

Bluetooth Low Energy in C++ for nRF Microcontrollers

Tony Gaitatzis, 2017

ISBN: 978-1-7751280-7-6

(This page intentionally left blank)

Made in the USA
Las Vegas, NV
04 November 2026

57430564R00057